CHAPTER – I

INTRODUCTION

1.1 OVER VIEW OF MATERIAL RECYCLING

Cars are one of the most recycled commercial products. Currently, approximately 75% of the total vehicle weight is recycled. The End-of-life vehicles try to push the recycling process further: it fixed the percentage of recyclability (85%) and recoverability (95%). The motivation for this work comes first from the observation of the material consumption of the automotive industry. The automobile is a major material consumer. Nowadays 95% of all vehicles go through the recycling process at their end of life. This collection rate is very high. By comparison, it is only of 52%, 55% and 42% for appliances, the efficiency of the recycling industry leads to a 75 % weight of vehicle recycled. Finally, the remaining 25%, known as Auto Shredder Residue (ASR), go to landfill. ASR is mainly composed of foams and fluff (40-52%), plastics (20-27%), rubbers (18-22%) and metals (4-15%) and there is currently no cost-effective recycling technology for plastics and foam.

The elements of the recycling chain which deal directly with End of Life Vehicles can be divided into three major steps: pretreatment, dismantling, and shredding. Emerging from each of these steps is a set of parts or materials which then pass to more specialized facilities for reprocessing. At its end-of-life, a vehicle typically goes first to the store yard of a dismantler, which can either be open and exposed to the elements or covered. The dismantler first removes all hazardous parts and fluids. This step is called pretreatment. The fluids typically removed include engine oil, coolant, refrigerant, steering oil, washer fluid, antifreeze, transmission oil, brake fluid, fuel, coolant and any remaining fuel. These fluids can either be removed by gravity or using pumps. At this stage, the dismantler typically also removes the tires, batteries, airbags and all parts presenting a potential hazard. After that, the hulk is typically crushed. It is then easy to handle and to transport to the next step: the shredder. The shredder takes the compacted car through hammer mills. These hammer mills shred the vehicle.

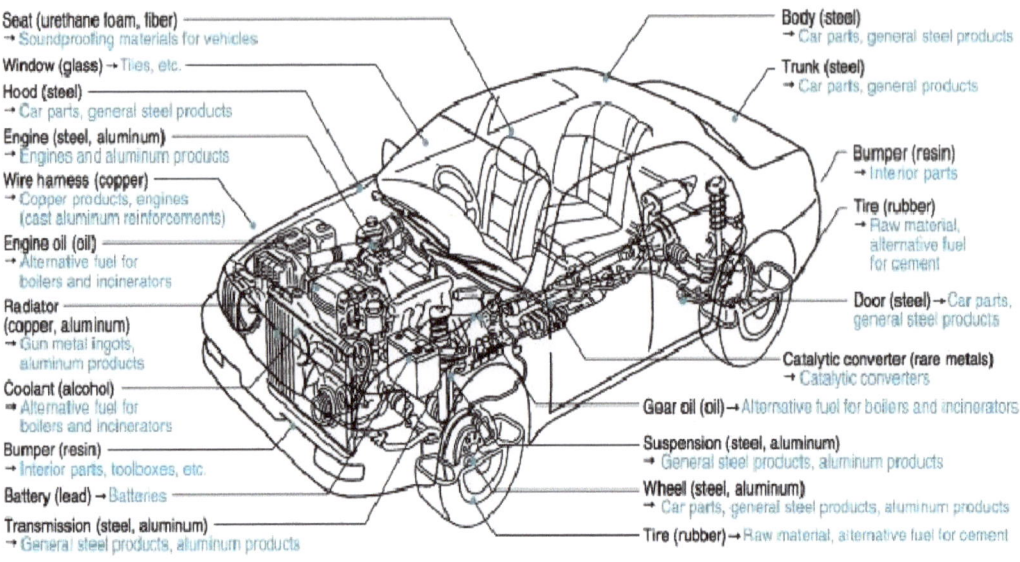

Fig. 1.1 Over view of Recycling Materials (adapted from: Toyota Company)

A well developed infrastructure exists for the reuse and recycling of automotive parts and materials. At the end of a vehicle's useful life many parts are removed and sold for reuse and fluids are recovered for recycling or proper disposal. What remains is shredded, along with other metal bearing scrap such as home appliances, demolition debris and process equipment, and the metals are separated out and recycled. The remainder of the vehicle materials is call shredder residue which ends up in the landfill. As energy and natural resources becomes more treasured, increased effort has been afforded to find ways to reduce energy consumption and minimize the use of our limited resources. Many of the materials found in shredder residue could be recovered and help offset the use of energy and material consumption. For example, the energy content of the plastics and rubbers currently land filled with the shredder residue is equivalent to 16 million barrels of oil per year.

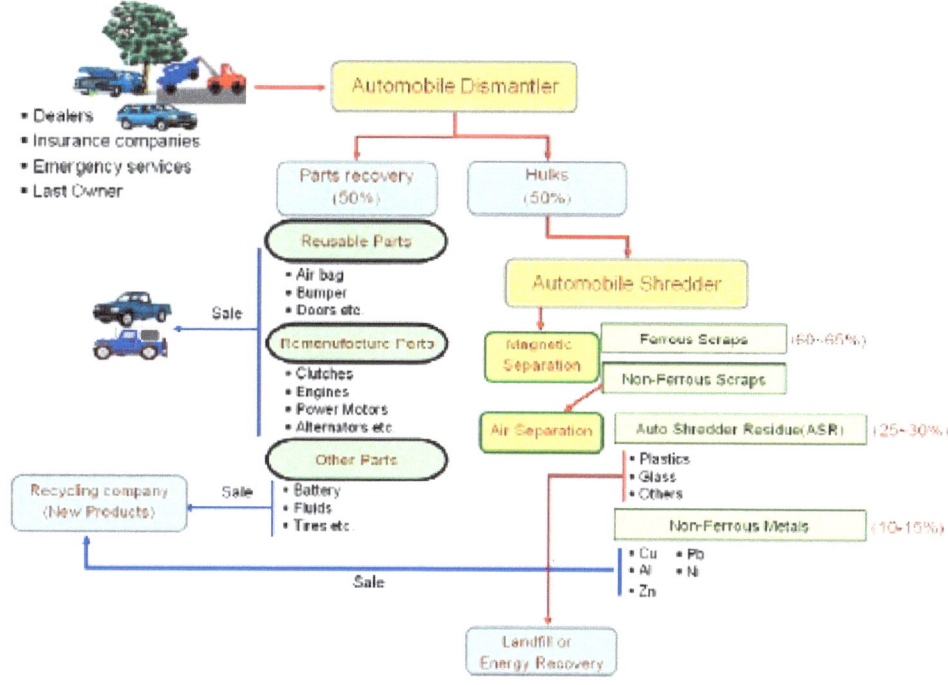

Fig. 1.2 Over view of Recycling Process (adapted from: Kim. J, et al, 2007)

1.2 MATERIAL SEPARATION PROCESS

1.2.1 DISASSEMBLY

Disassembly is the process of systematic removal of desirable constituent parts from an assembly while ensuring that there is no impairment of the parts due to the process. There are both economic and environmentally sound reasons for disassembly:

• *Discontinued products*: A suddenly discontinued product line can lead to excess inventory of undesirable assemblies. Disassembly scheduling can be used to retrieve valuable components (or components in short supply) which are common to other products still being produced. The remainder are recycled, sold or stored for future use.

• *Reduction in lead time*: Certain products might have to be disassembled in order to recover some of their subassemblies/components which are scarce and are in urgent demand by some other products. In such a situation, procuring the subassemblies/components needed for the urgent demand product(s) by disassembling other products may result in a substantial lead time reduction for these urgent products.

• *Forced disassembly*: A plant may be forced to disassemble inventories (before discarding) in order to comply with recycling regulations imposed by governments.

The most significant technical challenge lies in the design of the product. Designing a product with "easy" assembly constraints as well as "easy" disassembly constraints is likely to be a very difficult task. The current maturity level of product designs is mostly limited to assembly manufacturing. Disassembly is a new domain and the leading manufacturing companies have only recently realized its importance. In the past, products and machines were designed with only the assembly operations in mind. Now, designers have to think in terms of disassembly and parts recycling as well. Some of the design problems that need specific attention are:

• *Ease of separation*: Design for ease of separation, handling, and cleaning of all product components.

• *Low energy usage*: Design should aim at reducing energy usage for assembly as well as disassembly.

• *New fasteners*: New two-way snap-fit or pop-in pop-out fasteners should be developed, and existing ones should be improved. Screws, glues and welds should be replaced by other fastening methods. Taking apart a snap-fitted or pop-in pop-out product is much easier and requires less energy than taking apart a welded product.

• *Precision moulds*. Moulds should be designed to tighter tolerances so that the individual snap-fitted parts "stick" well to each another. Large tolerances with a snap-fitted product will reduce its lifetime and make it noisy when handled.

• *Materials selection*. The variety of material types has to be minimized in order to realize volume advantage from the building of large and efficient plants for recycling. Highly recyclable materials such as aluminum and thermoplastics should be encouraged, while minimizing the use of thermo sets which cannot be recycled.

• *Parts consolidation and product structure compression*. The product design should minimize the total number of stations and operations. Product structure compression is desired along with parts consolidation and commonality. This will allow easy sorting of the various components for recycling as well as easier assembly/disassembly operations involving fewer steps.

• *Technical problems with existing products.* Manufacturers may not be willing to redesign successful products completely and will only try to comply with disassembly constraints in their newer models. These modified designs for disassembly, which have not been initially conceptualized with disassembly in mind, may face some problems that affect their quality and reliability.

1.2.2 SHREDDING

After auto recyclers are finished dismantling a car and recovering parts, it is flattened and compressed, and directed to a shredder for scrap metal recovery. Along with ELVs, shredders process other metals-rich scrap, such as construction and demolition waste, and large end-of-life appliances (i.e. white goods). During the shredding process, the vehicle is broken down into much smaller pieces, and the metals are extracted. Both ferrous metals – iron and steel – and non-ferrous metals, such as copper, zinc and aluminum, are recovered. Ferrous metals make up about 70% of a vehicle, while non-ferrous metals make up about 6%.

The amount of recyclable material, then, that is removed from an end-of-life vehicle via shredding is generally calculated to be about 75% by weight. By far, the greatest percentage by weight of recycled material is the scrap metal. However, because this estimate of 75% does not take into account the parts, and materials recovered by dismantlers, it may be an underestimate of the percentage by weight of materials recovered for recycling and reuse. Therefore, it is possible that some auto recyclers are already recovering as much as 85% of end-of-life vehicles by weight. However, if we take 75% as a rough calculation of the weight of a vehicle that is currently being recycled, the residual 25%, which is left after metals are removed from the shredded vehicle, becomes waste and is sent to landfill. This 25%, known as shredder residue (SR), is generally a commingled mix of rubber, plastics, glass, dirt, carpet fibers, and seat foam, as shown below.

1.2.3 AUTOMOTIVE SHREDDER RESIDUE (ASR)

Generated ASR contains the bulk of non-metallic materials present in shredder hulks (plastics, glass, rubber, foam, carpeting, textiles, etc), entrained metallic fines; dirt and moisture. Two types of ASR streams can be generated from overall ELV processing:

• "Light" ASR ("fluff"): Generated at the shredder facility when the nonferrous fraction is separated into metal and nonmetallic streams using air classification processes (the nonmetallic fraction being "fluff").

• "Heavy" ASR: Generated at the non-ferrous metal processing facility during separation of the various metal steams (the heavy ASR representing rejected contaminants extracted during processing).

Currently, due to a variety of reasons, both light and heavy ASR is land filled "as is" in industrial landfills. Both types of ASR contain similar materials (plastics, glass, rubber, foam, carpeting, textiles, metallic fines, dirt, moisture, etc.), just in different proportions (light ASR containing a larger proportion of lighter materials like plastic and rubber; heavy ASR containing a larger proportion of heavier materials like glass and metal fines).

CHAPTER – 2

END OF LIFE VEHICLE MATERIALS

2.1 THE STATUS OF END OF LIFE VEHICLE MATERIALS

The ultimate goals for managing end-of-life vehicles should be to maximize the use of resources, to minimize pollution and to preserve and enhance employment. Producers and parts manufacturers have the greatest influence on the process that determines the waste management potential of a vehicle. To achieve these goals in an ideal world, producers would design vehicles in such a way that all parts in an end-of-life vehicle could be dismantled and either reused or recycled. Recycling materials has been shown to create 10 times more jobs and revenue than treating the materials as waste. The following table illustrates the approximate weight of the materials used in vehicles, and the trend lines showing which ones producers are choosing to increase or decrease. As the table suggests, lighter weight plastic components are replacing metal components in order to improve fuel efficiency in vehicles.

Ferrous and Non-Ferrous Metals: Approximately 76% of the weight of an average car is metal, primarily steel. Although the proportion of metal to plastics is decreasing, there is a very high recovery rate for metal from vehicles in all countries.

Plastics: The next largest component of vehicles by weight is plastic, which accounts for about 10% of the vehicle's weight. Currently, the amount of plastics being recycled is very low. Although the proportion of plastic being used in vehicles has increased, the proportion being recycled has not.

One reason for this is the variety of different polymers being used, which include polypropylene (PP), polyethylene (PE), polyurethane (PU) and polyvinyl chloride (PVC). Polypropylene accounts for the majority of car plastics (about 40%), and is used in bumpers, wheel arch liners and dashboards. Like polyethylene and polyurethane, which is common in seat foam, it is easily recycled. Markets for all of these polymers already exist. PVC, however, accounts for about 12% of the plastic used in vehicles, and is very difficult to recycle. No systems are in place for recycling PVC, and disposing of PVC by incineration releases dioxins.

One of the ways in which plastic recycling could be increased is through the identification of the different plastic polymers used in vehicle manufacturing. Marking components during production would make it possible for auto recyclers to remove and sort plastics in retired vehicles. The European's ELV Directive requires that member countries ensure that producers use material coding standards that allow the identification of various materials during dismantling.

	Kilogram per tonne of ELV		
Material/Fraction	2002	2006	2015
Ferrous Metal	680	680	650
Non Ferrous Metal	80	80	90
Plastics and Process Polymers	100	100	120
Tires	30	30	30
Glass	30	30	30
Batteries	13	13	13
Fluids	17	17	17
Textiles	10	10	10

Table 2.1 Composition of ELV Materials (GHK/Bio Study 2006)

Tires: Tire recycling programs have now been set up in all provinces. Tires represent about 3% of the weight of a vehicle. They are generally financed by an environmental fee charged on the purchase of new tires. Used tires are burned to produce fuel, or recycled into a variety of products, including automotive products. Fine mesh crumb or ground rubber can be used as protective liners for truck boxes or as ingredients in new tires. An (European) EU landfill ban on tires increased the percentage of tires being recovered in European countries from 60% to over 95%.

Glass - Used windshield glass, which is 3% of the vehicle weight, has been a very low priority for recycling. As a result, an estimated 18 to 20 million kilograms per year of used glass is disposed of in landfill sites.

There are two types of windshield glass – toughened and laminated. While it is easy to remove toughened glass when it is shattered, laminated glass does not shatter and needs to be removed manually. Laminated glass is two layers of glass with a strong plastic membrane between the panes. It can only be recycled if the plastic film is separated from the windshield. It can be turned into construction aggregate, Glasphalt or secondary markets like floor tile if the glass can be separated from the plastic film. However, these processes are expensive and not common. In addition to the challenges of removing and recycling laminated glass, windshield glass is not generally recycled because its value is relatively low, and there are no financial incentives. If glass is recovered, 1.2 tons of raw materials are saved for every tonns of crushed glass used in manufacturing.

Batteries – Recycling programs for lead-acid batteries are in place across Canada, and it is estimated that more than 90% are recycled. Industry results from US data show that about 97% of spent battery lead was recycled from 1997 to 2001. Although batteries comprise only about 1% of a vehicle by weight, it is important that they are recovered and recycled because lead is hazardous.

Fluids: All provinces have some form of stewardship program to recover and recycle used motor oil and other lubricants.

Textiles: Textiles refer to carpets and other upholstery in ELVs. They make up about 1% of the vehicle by weight, and are usually disposed of as auto shredder residue.

Rubber: – Similarly, rubber, which comprises 2% of the vehicle by weight, is not recovered, and goes to landfill as part of the auto shredder residue.

2.2 THE ELIMINATION OF HAZARDOUS CHEMICALS

The accelerating change in materials composition (for example the increasing of the fraction of plastic and aluminum) of modern vehicle can create new problems in the recycling process of ELV. Recycling of plastic is very difficult when it is present in small parts or attached to another material. Similarly, recycling of aluminum is not straightforward because it is normally present in the form of alloys. Based on that scenario, material choice is one of the key elements in vehicle design in order to make the concept of ELV successfully implemented. This is basically because the different materials have a different technique of disassembly and recycling. Plastic contributes around 9% of the weight of an ELV and this is increasing as vehicle manufacturers continue to develop lightweight vehicles to improve fuel efficiency (DTI Report, 2003). The recycling rate of plastics needs to be improved because most of the plastic material from an ELV arises at the shredder as shredder fluff. Furthermore, the plastic materials are very difficult to extract for recycling unless they can be removed prior to shredding but this normally is costly unless easy removal is part of the original design.

Beyond the immediate potential for recycling vehicles on the market today, the goal of maximizing resources and attaining higher recycling rates can only be achieved in the design stage. An important aspect of designing for less environmental impact and greater recycling potential is minimizing the use of toxic chemicals in vehicle manufacturing. The use of less toxic chemicals would also reduce the risks to the health of workers in vehicle and parts manufacturing plants, workers in the auto recycling business, and people in their vehicles. For example, the elimination of polyvinyl chloride (PVC) in vehicle components would reduce workers' exposures to a chemical with possible carcinogenic effects. As mentioned before, the EU's ELV Directive promotes the reduction and elimination of hazardous chemicals.

It encourages vehicle manufacturers to limit the use of hazardous substances in vehicles, particularly in the design stage, so that recycling is easier and hazardous substances are not released into the environment. The Directive also specifically targets the reduction or Elimination of lead, mercury, cadmium and hexavalent chromium. Although there are some exemptions, countries were responsible for ensuring that vehicles on the market after July 1, 2003 did not contain certain materials or parts containing these chemicals. In addition to heavy metals, environmental organizations have raised concerns about the use of certain plastics, brominates flame retardants and volatile organic compounds in vehicles. Many plastics, such as PVC used in dashboards and exterior trim, release toxic chemicals during production, use and disposal, as well as being difficult to recycle. In addition, PVC releases phthalates, another chemical of concern.

2.3 DESIGN FOR RECYCLABILITY

Design-for-Recyclability (DfR) is a product design tool that considers the materials from which a product is manufactured and how these materials are assembled. If applied during a product's conception and carried through to its design, assembly, and ultimately disposal, these criteria can be an effective tool to minimize wastes and maximize the reuse of materials.

General DfR criteria applied to automotive design for end-of-life recyclability include:

Use recyclable materials:

Design products using materials that can be recycled and for which materials collection and recycling technologies currently are available and commonly used. Generally, metals are easier to recycle than nonmetals, and thermoplastic resins are more desirable than thermo set plastics. Alternatively, set up an effective materials collection system (e.g., offer to accept used lead-acid batteries when new batteries are purchased).

Use recycled materials:

Select materials that contain a high percentage of recycled content, as this supports the recycling process for which a product is being designed. Steel and aluminum are materials that are often recycled.

Reduce the number of different materials used within an assembly:

Reduce the number of materials used to manufacture a component or assembly. Reducing the number of materials also simplifies the separation process and supports recycling.

Mark parts for simple material identification:

Mark all materials with standard material identification codes. Although this process is most feasible for plastic parts, it can be expanded to metals, composite materials, and coatings currently used in vehicle manufacturing.

Use compatible materials within an assembly:

Select materials that do not need to be separated for recycling. Generally, mixtures of dissimilar plastics cannot be recycled. Similarly, nonferrous metals (e.g., aluminum, chromium, or zinc) can contaminate and thus decrease the recyclability of ferrous metals (i.e., iron and steel), and vice versa. Layers of paint or plated metal over a base material also represent contaminants not compatible with recycling. If a coating on metal cannot be removed, the paint or metal plating will be a contaminant that decreases the metal's recyclability and/or the applications for which the recycled metal can be used.

Make it easy to disassemble:

Also called Design for Disassembly, this criterion guides a designer away from complicated products and assembly processes. Using snap fits and nut/bolt assembly techniques whenever possible assists in disassembly, as does avoiding adhesives, particularly when bonding two incompatible materials or if the adhesive will contaminate the materials so they cannot be recycled.

CHAPTER – 3

END OF LIFE VEHICLE MANAGEMENT

3.1 End-of-Life Vehicle Management:

The ELV management system lacks regulations and it's controls however, governments are developing guidelines for automotive recyclers to follow. Governments tend to approach the ELV management systems using an EPR (Extended Producer Responsibility) approach. EPR makes the original manufacturer responsible for the end-of-life management costs by doing this it promotes manufacturers to increase the recyclability, reduce the wastes and toxicity, and allow for easier reuse and remanufacturing of their products. While some governments opt for full EPR, others tend to lean towards shared responsibility between the manufacturers and government.

The ELV management activities are:

1. Dismantlers, consisting of two distinct types:

   - High-value parts dismantlers (high volume, quick turnover operations targeting late-model vehicles).

   - Salvage/scrap yards (low volume, slow turnover operations accepting most vehicles).

2. Shredding facilities.

3. Non-ferrous separation facilities

4. Steel mills (specifically, Electric Arc Furnaces –EAFs)

5. Landfills.

Automobile owners permanently retire their vehicles for a variety of reasons such as:

- Loss of structural/mechanical integrity from corrosion or an accident

- Poor reliability of parts and components

Retired Vehicles (End-of-Life Vehicles and Premature End-of-Life Vehicles from Accidents)	
Pretreatment (remove battery, fluids, tries, mercury switches, ozone-depleting substances, airbags)	Reuse (batteries, fuel, fluids such as antifreeze and windshield washing fluid, tires, ozone-depleting substances) Recycling (batteries, fluids, tires, ozone-depleting substances)
Dismantling (remove parts/,materials)	Reuse of parts Remanufacturing of parts Recycling of materials
Shredding (shred vehicle, reclaim metal)	Recovery of metals
Landfill Shredder Residue	

Table 3.1 Process of End-of-life Vehicles.

If a vehicle is taken to an auto recycler, the parts that can be recycled or used to replace parts in other vehicles will be removed before the vehicle is sent to a shredding operation. Parts that can be recovered for reuse or remanufacturing include AC compressors, water pumps, carburetors, calipers, power steering pumps, carrier assembly, alternator, starters, transmissions, axle assemblies, engines, and transfer cases. Batteries, catalytic converters, radiators and tires are also removed from end-of-life vehicles for recycling.

3.2 GOALS OF ELV DIRECTIVE

The directive covers aspects along the life cycle of a vehicle as well as aspects related to treatment operations. As such it aims at

- preventing the use of certain heavy metals such as cadmium, lead, mercury and hexavalent chromium,

- collection of vehicles at suitable treatment facilities,

- de-pollution of fluids and specific components,

- coding and/or information on parts and components

- ensuring information for consumers and treatment organizations

- achieving reuse, recycling and recovery performance targets

In recent years, environmental issues have become a priority for manufacturing companies. In particular, the automotive industry has taken a proactive stance due to legislative pressures. Legislation such as the End-of-Life Vehicle (ELV) Directive (The European Parliament and of the Council of European Union, 2000, 2002, 2005a, 2005b, 2005c, and 2008) has highlighted the need for automotive Original Equipment Manufacturers (OEMs) to design vehicles that can conform or, indeed, exceed ELV targets. At present, approximately 75% to 80% of end-of-life vehicles in terms of weight, mostly metallic fractions, both ferrous and non ferrous are being recycled. However, the remaining 20% to 25% in weight, consisting mainly of heterogeneous mix of materials such as resins, rubber, glass, textile, etc., is still being discarded (Toyota Motor Company, 2005).

EU ELV Directive forces the vehicle manufacturers to (The European Parliament and of the Council of European Union, 2000, 2002, 2005a, 2005b, 2005c, 2008):

1) Reduce the use of hazardous substances.

2) Design new vehicles that are easier to dismantle, reuse, recycle and recover components/materials/energy from vehicles that have been junked or totaled.

3) Increase the use of recycled materials in new vehicles.

The EU draft on ELVs also outlined that car manufacturers must reuse or recover 85% of ELV by 2006. Stating that at least 80% of a vehicle's weight must be reused or recycled; although up to 5% can be dealt with through other recovery operations such as incineration. This target increases to 95% by 2015 and at least 85% of that weight must be reused or recycled (The European Parliament and of the Council of European Union, 2000, 2002, 2005a, 2005b, 2005c, 2008). A summary of the general recycling targets, based on the ELV Directive, and recycling targets for the type-approval of new vehicles are shown in Table 3.2.

Year	Event
2000	EU Directive on ELV was signed by the European Parliament and Council of Ministers
2002	Free of charge take back of new cars
2003	Use of certain heavy metals forbidden: Cd,Cr(VI),Hg.Pb.
2005	Types of approval: OEMs have to prove that car meets 2015 recycling/recovery quotas
2006	Dismantlers have to meet following quotas: $\geq 80\%$ recycling, $\leq 10\%$ energy recovery, $\leq 15\%$ landfill.
2007	Free of charge take back of all ELVs
2015	Dismantlers have to meet following quotas: $\geq 85\%$ recycling, $\leq 5\%$ energy recovery, $\leq 5\%$ landfill.

Table 3.2 Summary of the ELV Directive.

3.3 ELV MANAGEMENT IN THE UNITED STATES, EUROPEAN UNION, JAPAN AND SOUTH KOREA

United States

- The management of ELVs involves dismantling, shredding, and recycling of parts and materials. The material recycling rate in the United States is 75% by weight. The remaining 25% is composed mainly of automotive shredder residue (ASR) and fluff.

- Approximately 2.5 to 3.0 million tons of ASR are produced and disposed of each year. Currently, 94% of cars in the United States are sent to dismantling and shredding facilities at the end of their service life.

- There are no strict regulations regarding land filling of ASR in the United States, as ASR is considered a non-hazardous waste.

- Only California has classified ASR as a hazardous waste, resulting in high management costs for ASR disposal.

- One major project of the VRP (vehicle recovery process) is the Vehicle Recycling Development Center, established in 1993 as the first Big Three joint research facility.

- The VRP is working with the American Plastics Council on developing pyrolysis technology to decompose plastic wastes to a hydrocarbon gas and oil that can be used as a feedstock to produce new plastics.

European Union

- The publication of Directive 2000/53/CE of the European Parliament and the Council of 18 September 2000 on ELVs imposed detailed targets for ELV management. It required reuse and recovery rates of 85% by 1 January 2006 and 95% by 1 January 2015; it also stipulated a reuse and recycling rate of 80% by 15 January 2006 and 85% by 1 January 2015.

- The Directive also established Extended Producer Responsibility (EPR) for ELV management. The EPR requires manufacturers and importers of automobiles to pay for the end-of-life costs associated with recycling. Beginning January 1, 2007, they are responsible for the recycling costs of all vehicles, regardless of age.

- Other provisions of the Directive include Design for Environment (DfE) practices, increased quantity of recycled materials in automobiles, component and material coding for product identification and dismantling information for every vehicle.

- Finally, the Directive requires that all vehicles put on the market after July 1, 2003 contain no lead, mercury, cadmium, or hexavalent chromium, except in certain excluded components (e.g. lead in lead-acid batteries, hexavalent chromium as a corrosion preventative coating, lead-containing alloys of steel, aluminum and copper and mercury in headlamps).

Japan

- The number of ELVs in Japan is estimated at 5 million units per year. The material recovery rate is about 75% and the amount of (landfilled) ASR is estimated at 800,000 tons per year. The ELV treatment operations are performed by an estimated number of 3,500-5,000 dismantlers. The dismantling sector is not considered to be well-organized and inappropriate treatment is assumed to result in adverse environmental impacts.

- Japan has no specific regulation on ELVs but various laws create a framework for ELV management. The Environmental Law of 1994 includes the objectives for waste reduction, reuse of end-of-life products, promotion of recovery, recycling and appropriate waste processing. The Waste Disposal Law, revised in 1997, introduced heavier sanctions on inappropriate waste management and created additional tasks for regional governments.

- The development of plans by individual manufacturers started in 1998. They are based on the quantified targets described above and include: (a) technical developments on dismantling; (b) assistance to dismantlers; (c) preparation of dismantling manuals; (d) research on reuse of parts from ELVs; (e) technical research on energy recovery from ASR.

South Korea

- The management of vehicles at the end of their service life is conducted by vehicle recycling industries in South Korea. Except for parts restricted by law, reusable automobile parts are dismantled and then distributed by most recycling industries through the Internet for reuse.

- A policy for ELVs management in South Korea was first established in December 1982 as the Road Transportation Vehicle Directive, now known as the Vehicle Management Directive. Under this directive, the final owner of an ELV has the responsibility of delivering it to the auto dismantler.

- Europe and Japan have similar ELVs management systems where ELVs certificates are needed for cancellation of registration.

- The Korean scrap vehicle industry was established in 1982 and, in order to broaden its operations and vehicle management capabilities, it was officially acknowledged as a commercial sector in 1995. In 1995, there were 141 scrap industries in South Korea, while in 2003; the number has increased to 310. Within a period of 8 years, the number of scrap industries has become 2.2 times larger.

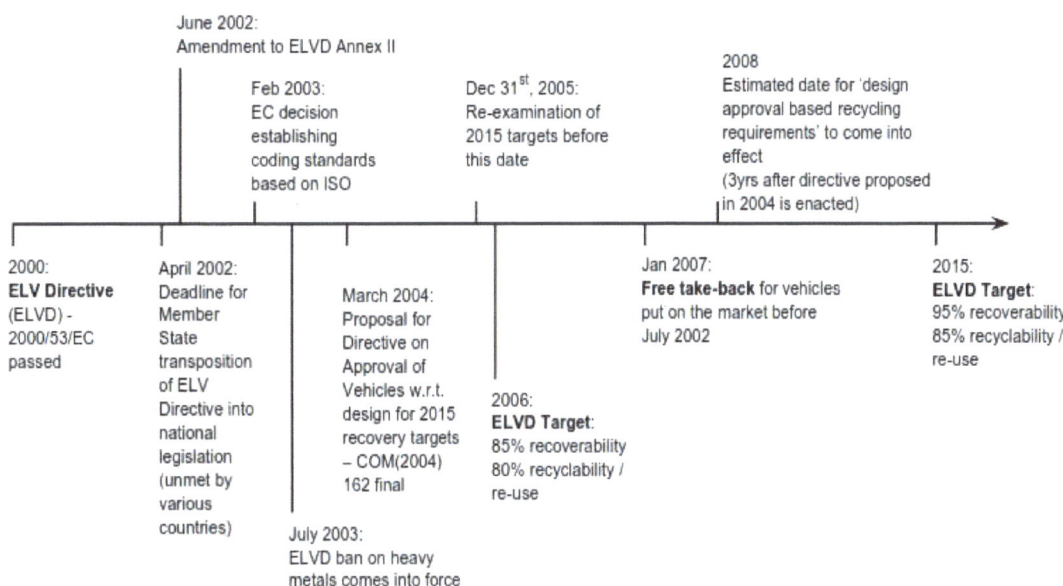

Figure 3.1 ELV legislation timeline (Adapted from: EU Directive)

3.4 RECYCLING MODEL PLAN:

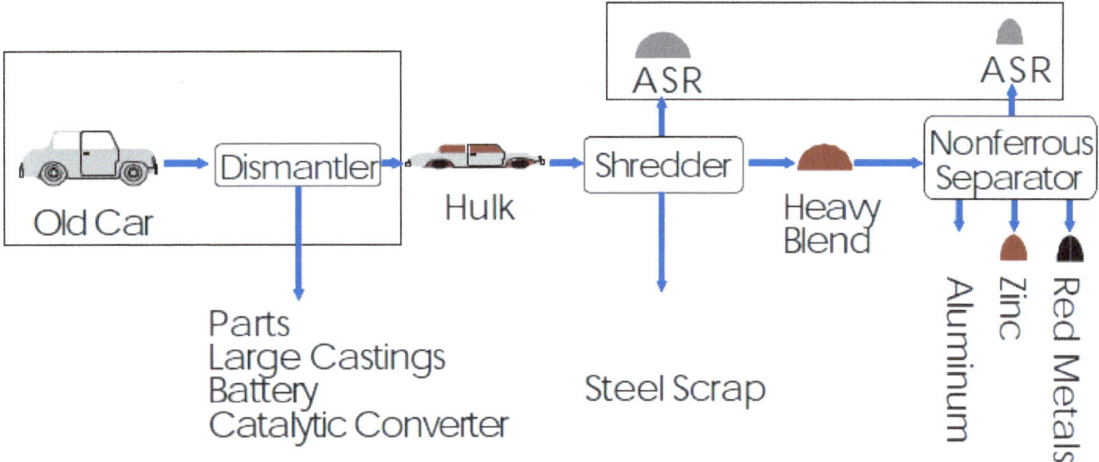

Figure 1.3 Recycling Model Plan (adapted from: Bae, J. and Kim, J. 2006)

CHAPTER – 4

LITERATURE REVIEW

A **literature review** is a text written by someone to consider the critical points of current knowledge including substantive findings as well as theoretical and methodological contributions to a particular topic. Literature reviews are secondary sources, and as such, do not report any new or original experimental work. Also, a literature review can be interpreted as a review of an abstract accomplishment. Most often associated with academic-oriented literature, such as a thesis, a literature review usually precedes a research proposal and results section. Its main goals are to situate the current study within the body of literature and to provide context for the particular reader.

"The Green Degree Assessment of the Automobile Reverse Logistics System", in this paper Meng Xiangru and Song wei use the fuzzy comprehensive analysis method to judge the green degree of the automobile of reverse logistics to builds up a green degree factor sets. By improving the software and hardware facilities to achieve win-win economic and social benefits, improve the overall level of functioning of reverse logistics system.

"Inverse Factory for End-of-Life Cars-Complete Dismantling System", in this paper Hironobu Yamamoto, Susumu Shibata and Henk C. Neijenhuis, present a new concept of recycling system that is a complete dismantling of end-of-life-cars that goes along with human oriented working environment. (i.e.) Complete Dismantling of EOL cars. This system gives the solution for the car shredder dust problem.

"Automotive Green Supply Chain Management Based on the RFID Technology", in this paper Tongzhu Zhang, Xueping Wang, Xianghai and Jiangweei Chu analyzes the requirements of EPR legislation, automotive supply chain, RFID (Radio Frequency identification devices) and information tracking system, and then puts forward the information flow mode and operation process of information management system based on the supply chain using RFID technology. It can help the producers to establish a green supply chain management system and finally to achieve the goal of environmentally friendly vehicle and high recycling/recovery rate of their ELVs.

"Graph-Based Information Modeling of Product-Process Interactions for Disassembly and Recycle Planning" in this paper Satoshi kanai, Ryohta Sasaki, and Takeshi Kishinami proposes an information modeling to represent the product and the process and their interactions in the disassembly and recycle planning of consumer product. They used four kinds of graph to express the models also gives a software tool for planning all feasible disassembly and recycle process.

"Dynamic Speed Flexible Drive System for Shredder-Plants with Highly Restricted Control Range" here Sourkounis tells that Shredder drives gives required Mechanical power and this process are affected by stochastically load with high amplitudes in electromechanical drive train. Based on the operating performance of commercial shredder plants, a new drive train with a novel non-linear speed control was developed and it realizes a rotational speed-flexible operation. In this way it was possible to reduce the cumulative load in the drive train, achieving high process performance.

"An Information System Management of Assessment of Disassembly and Recycle" in this paper Meng Peng, DUAN Guanghong, and XIANG Dong presents an adaptive information management system for product planning and evaluation during the design phase for disassembly and recycling. The framework of the proposed information management system for modeling and assessment is demonstrated comprehensively in disassembly of air-conditioners.

"Automotive Recycling Information Management Based on the Internet of Things and RFID Technology" in this paper Tongzhu Zhang, Xueping Wang, Xianghai and Jiangweei Chu points out the importance and current development of automotive recycling industry in China and introduces the internet of things, RFID (Radio Frequency identification devices) technology with their applications. They give a basic model of recycling information management system and information flow at each stage of a vehicle lifecycle (beginning, middle and EOL), hoping to promote the development of automotive products recycling industry in China.

"Optimizing Disassembly and Recycling Process for EOL LCD-Type Products: A Heuristic Method" in this paper Li-Hsing Shih and Shun-Chung Lee proposes a two-stage heuristic method to simulate and identify an optimal disassembly sequence as well as a break point between the disassembly and shredding processes. This method is used to find an optimal EOL recycling process and will not only help recyclers plan but also allow manufacturers outside the EU to anticipate how the EOL process will work and estimate the costs.

"Conceptualizing an Optimal Electronic Product Design and End-of-Life Management System" in this paper Wayne Rifer, Jeff Omelchuck, John Katz and Viccy Salazar identifies the importance of electronic product reuse, dismantling and component recovery for an environmentally preferable end-of-life management system, and concludes by noting that in the US the recycling infrastructure is still largely dependent for its economic sustainability on reuse.

"Study on the Informationization Management for Auto-Remanufacturing Recovery System" in this paper Shao Zhifang studies on the remanufacturing system's characteristics, the main elements involved and the data message's characteristics, provide a main information management system framework for auto-remanufacturing.

"Recovering Copper from Spent Lithium ion Battery by a Mechanical Separation Process" in this paper Shuguang Zhu, Wenzhi Hel, Guangming Lil, XuZhoul, Juwen Huang, and Xiaojun Zhang explains that based on the structure of lithium-ion batteries (LIBs), the electrode materials were separated from spent LIBs with aim to recycle all valuable components as possible. After shredding and sieving most copper was concentrated in the particle size above 0.59mm, the copper recovery rate reached 93.10 wt %, and the content of copper was 95.40% at the condition of 3 min pulverization. Further separation of the anode scraps from 0.590 mm to 0.177 mm was carried out using fluidized bed technology. Approximately 92.30 wt % of copper in anode particles from 0.590 mm to 0.177 mm can be recovered by a gas-fluidized bed separator at the selected optimal gas velocity (1.00 m·s-I).

"Evaluation of Value Changes Between Different Phases of the Product Life-Cycle" in this paper George L. Kovacs says that Product Life-Cycle Management (PLCM) starts when the idea of a product is born and lasts until complete dismissal through steps as requirements, analysis, design, part manufacturing, assembly, testing etc., His real goal is to give some means and tools to calculate different values, expressed in money, which correspond to different phases of a product life-cycle (PLC). They plan to find proper relationships to use their ideas and formulae for real world situations to assist not only designers and engineers in their work, but politicians and other decision makers as well.

"Using Fuzzy Cognitive Map for Evaluation of RFID-based Reverse Logistics Services" in this paper Amy J. C Trappey, Chang-Ru Wu, Fu-Chiang Hsu and Charles V.

Trappey proposes a decision support model that integrates fuzzy cognitive maps trained using a genetic algorithm. The advantage of using fuzzy cognitive maps is that the model and the relationships among nodes can be expressed both quantitatively and qualitatively and also to diminish the subjective effects of the weights, the genetic algorithm is applied. The aim of this paper is to propose a FCM model for decision makers to better understand the outcomes of reverse logistics processes.

"Is European end-of-life vehicle legislation living up to expectations? Assessing the impact of the ELV Directive on 'green' innovation and vehicle recovery" in this paper Gerrand and Kandlikar presents an evaluation framework based on five anticipated changes that could result from the ELV Directive. These changes relate to three areas: (a) vehicle design, (b) level of ELV recovery, and (c) information provision. We evaluate the extent to which expected outcomes have materialized since the establishment of the ELV Directive. Increasing the level of re-use and remanufacturing will be a key part of moving toward sustainable vehicle production.

"End of life vehicles recovery: process description its impact and direction of research" in this paper Muhammad Zameri b. Mat Saman presents current practices in vehicle recovery in EU, USA. The concepts of sustainable development and end of life vehicle recovery, and the long terms changes required to more readily support the core themes of the end of life vehicle recovery.

"Life Cycle Assessment on End of Life Vehicle Treatment System in Korea" in this paper Kee Mo Jeong, Seok Jin Hong, Ji Yong Lee, and Tak Hur aim is at evaluating the environmental impacts stemmed from End of Life Vehicle treatment system using Life Cycle Assessment method.

"A Design Framework for End of Life Vehicle Recovery" in this paper Chris Edwards, Tracy Bhamra, Shahin Rahimifard says that the Current vehicle design does not give a sufficient aid and the economic recovery of parts and materials to reach this target. Their aim is to provide a framework so that the cost of recovery can influence design.

"Life Cycle Assessment in the Automotive Industry: Comparison Between Aluminum and Cast Iron Cylinder Blocks" in this paper F. Bonollo, G. cupito and R. Molina demonstrates that while during the production stage the environmental load related to the aluminum block is higher than the one related to the cast iron block during use and end of life treatment the gain of aluminum over cast iron makes the aluminum cylinder block more environment friendly than the cast iron one. This paper presents the results of the comparison between the environmental load of cast iron and aluminum cylinder blocks. The methodological approach adopted for the analysis is Life Cycle Assessment (LCA) since it allows considering the environmental effects of a product during the production, use and end-of-life treatment phases.

"Life Cycle Assessment Practices: Bench marking Selected European Automobile manufacturers" in this paper Jean Jacques chanaron says that LCA has become a widely used set of tools for the management of all impacts on environment by industrial products. LCA is carried out at the very early stages of product research, development and design.

"Sustainability of the automotive recycling infrastructure: review of current research and identification of future challenges" in this paper Vishesh Kumar and John focuses on describing past research relative to the automotive recovery infrastructure and those research challenges that may arise in the future. In addition to the development of improved models, several questions, e.g., energy issues, life cycle CO2 emissions, effect of light weight materials, and impact of new power train technologies should be examined via scenarios that are require to better understand the new challenges.

"New methods for recycling plastic materials from end of life vehicles" in this paper Denis Panaitescu, Michaela Iorga, Adriana Ciucu, Sever Serban, Augustin Crucean, Cristiana Bercu presents some experimental results for recovering plastic materials from bumpers and new materials with Interesting properties were obtained by compounding these samples with wood flower and virgin polymers. Different samples obtained from bumpers were analyzed and mechanical characterized for the identification of polymers and their level of mechanical strength and thermal stability.

"Strategic Guidance Model for Product Development in Relation with Recycling Aspects for Automotive Products" in this paper Muhamad Zameri Mat Saman discusses a strategic guidance model for the product development process of automotive components in order to fulfill the requirements of the recycling aspects in End-of-Life Vehicle (ELV) Directive. This paper also presents an example of the whole vehicle as a case study in order to demonstrate and validate the proposed framework.

"Technologies for the Identification, Separation and Recycling of Automotive Plastics" in this paper Joerg Hendrix, Kevin A. Massey, Eric Whitham, and Bert Bras provides an overview of efforts and technologies which primarily support automated separation and recycling. Although this paper is focused on automotive applications, many of the technologies are applicable to white goods and consumer electronic products as well.

4.1 SUMMARY OF LITERATURE REVIEW:

S.NO	TITLE AND AUTHOR NAME	ABSTRACT	TOOLS AND METHODOLOGY	JOURNAL NAME AND YEAR
1.	An Information System Management of Assessment of Disassembly and Recycle by **Meng Peng, DUAN Guanghong, and XIANG Dong**	This presents an adaptive information management system for product planning and evaluation during the design phase for disassembly and recycling.	Framework modeling	IEEE 2003. Q-7803-7743-5.
2.	Automotive Recycling Information Management Based on the Internet of Things and RFID Technology By **Tongzhu Zhang, Xueping Wang, Xianghai and Jiangweei Chu**	This points out the importance and current development of automotive recycling industry in China and introduces the internet of things, RFID technology with their applications.	Radio frequenting identification technology	IEEE 2010. 978-1-4244-6932.
3.	Conceptualizing an Optimal Electronic Product Design and End-of-Life Management System by **Rifer, Jeff Omelchuck, John Katz and Viccy Salazar**	This identifies the importance of electronic product reuse, dismantling and component recovery for an environmentally preferable end-of-life management system.	The Electronic Product Environmental assessment Tool	IEEE 2007.1-4244=0861-X.
4.	Study on the Informationizatio n Management for Auto-Remanufacturing	This studies on the remanufacturing system's characteristics, the main elements involved	Study project	IEEE 2010.

	Recovery System By **Shao Zhifang**	and the data message's characteristics; provide a main information management system framework for auto-remanufacturing.		978-1-4244-6932-1.
5.	A Design Framework for End Of Life Vehicle Recovery. By **Chris Edwards, Tracy Bhamra, Shahin Rahimifard**	Current vehicle design does not sufficient aid the economic recovery of parts and materials to reach this target. This paper aims to provide a framework so that the cost of recovery can influence design.	Study project	CIRP International Conference on Life Cycle Engineering. (2006)
6.	Environmentally responsive supply chains Learning's from the Indian auto sector. By **Apratul Chandra Shukla, S.G. Deshmukh and Arun Kanda**	This paper identifies the implementation level, major drivers, various practices and performance of environmentally and socially-conscious supply chain management (SCM) in the context of the automobile industry in India.	Supply chain management.	Journal of Advances in Management Research Vol. 6 No 2, 2009 pp. 154-171
7.	Sustainability of the automotive recycling infrastructure: review of current research and identification of future challenges by **vishesh kumar and john. w**	This paper focuses on describing past research relative to the automotive recovery infrastructure and those research challenges that may arise in the future.	Study project	International journal of sustainable manufacturing .(2008)
8.	New methods for recycling plastic materials from end of life vehicles By **Denis Panaitescu,**	Some Experimental results for recovering plastic materials from bumpers are presented in this paper. New materials with Interesting properties were obtained by	Analyzed by FTIR.	ISSN (2011)

	Michaela Iorga, Adriana Ciucu, Sever Serban, Augustin Crucean, Cristiana Bercu	compounding these samples with wood flower and virgin polymers.		
9.	Remanufacturing in Automotive Industry: Challenges and Limitations by **Paulina Golinska, Arkadiusz Kawa**	This paper presents the framework for management of reverse flow of materials on automotive industry.	Study project	Journal of Industrial Engineering and Management pp. 453-466 (2011)
10.	Comparative analysis of scrap car recycling management policies by **Kun Yue**	This paper uses GREET (The Greenhouse Gases, Regulated Emissions, and Energy Use in Transportation Model) to calculate energy use and GHG emissions in different ways of car recycling.	The Greenhouse Gases, Regulated Emissions, and Energy Use in Transportation Model	Procedia Environmental Sciences 16 (2012) pp. 44 – 50.

CHAPTER – 5

PROBLEM DESCRIPTION

In this project, we are tried to reduce the landfills of automobile recycling. For that, we create the new model for material recycling. After that, for analyzing we collect some data's from the previous work relating to recycling and we use the MINITAB software techniques and we the find the new Accelerated Life Testing for automobile recycling.

CHAPTER – 6

EXPERIMENTAL DESIGN

6.1 New Recycling Model

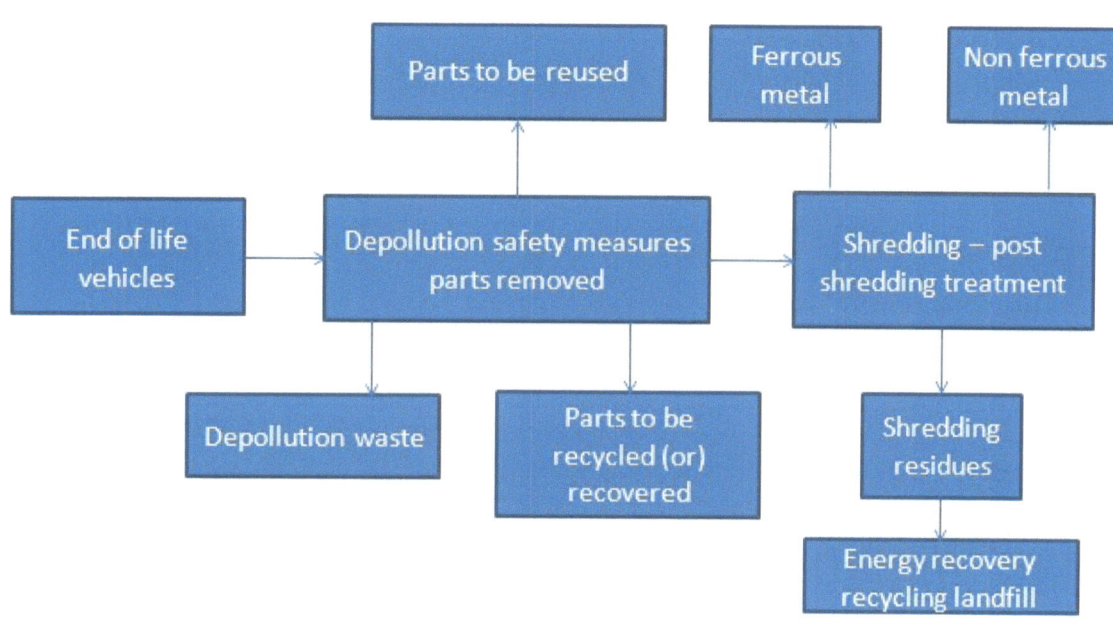

The above new recycling model is one of the effective recycling techniques for the automobile material recycling which reduces the scarab rate in the shredder residue. At present, modern recycling facilities are able to recover over 95% of the ferrous and nonferrous metals in ELVs. The less than 5% of the residual metals that remain in the shredder residue constitute about 5–15% of the weight of the shredder residue. To facilitate the recycling of non-metallic materials from shredder residue, the dismantling industry, repair shops, the shredding industry, and automobile manufacturers must work together.

6.2 Minitab Software

Minitab is a statistics package. It was developed at the Pennsylvania State University by researchers Barbara F. Ryan, Thomas A. Ryan, Jr., and Brian L. Joiner in 1972.

• Minitab is a powerful statistical software package that removes much of the pain associated with analyzing data and using statistical tools

• This module provides an introduction and overview of its use.

6.3 Use of Minitab

- Data and File Management - spreadsheet for better data analysis.
- Regression Analysis
- Power and Sample Size
- Tables and Graphs
- Multivariate Analysis - includes factor analysis, cluster analysis, correspondence analysis, etc.
- Nonparametric - various tests including sign test, runs test, Friedman test, etc.
- Time Series and Forecasting- tools that help show trends in data as well as predicting future values. Time series plots, exponential smoothing, trend analysis.
- Statistical Process Control
- Measurement System Analysis
- Analysis of Variance - to determine the difference between data points.

In this work the MINITAB 16 Version Software was used.

6.4 Accelerated Life Test

An accelerated life test models product performance (usually failure times) at elevated stress levels so that you can extrapolate the results back to normal conditions. The goal of an accelerated life test is to speed up the failure process to obtain timely information about products with a long life.

Use accelerated life test plans to determine the number of units to test and the allocation of those test units across stress levels for an accelerated life test. You can also use these test plans to determine the standard error for the parameter you want to estimate for a fixed number of test units.

Use accelerated life test plans to answer questions such as:

- How many units must I test to estimate the 10th percentile with a 95% upper confidence bound within 100 hours of the estimate?

- What is the best allocation of 20 units across 3 stress levels in order to estimate the reliability at 1000 hours?

- If twenty units are available for testing, what standard error can I expect for the estimate of the 500-hour reliability?

To obtain an accelerated test plan, you provide the stress values. In addition, you can provide the proportionate allocation of test units. Minitab evaluates the resulting plans and displays the "best" plans with respect to minimizing the variance.

6.5 Table of Input Parameters:

FERROUS	AVERAGE WEIGHT	NON FERROUS	SHREDDER RESIDUE	POLYMER CONCENTRATE
776	1302	194	512	347
645	1028	176	467	428
623	1154	213	450	490
547	1098	263	320	538
498	1275	346	298	587

6.7 Steps by Step Procedure for Accelerated Life Testing

First to open the MINITAB Software by using of icons.

Step 1: From the main MENU bar stat→ sat→ Reliability/Survival→ Accelerated Life Testing.

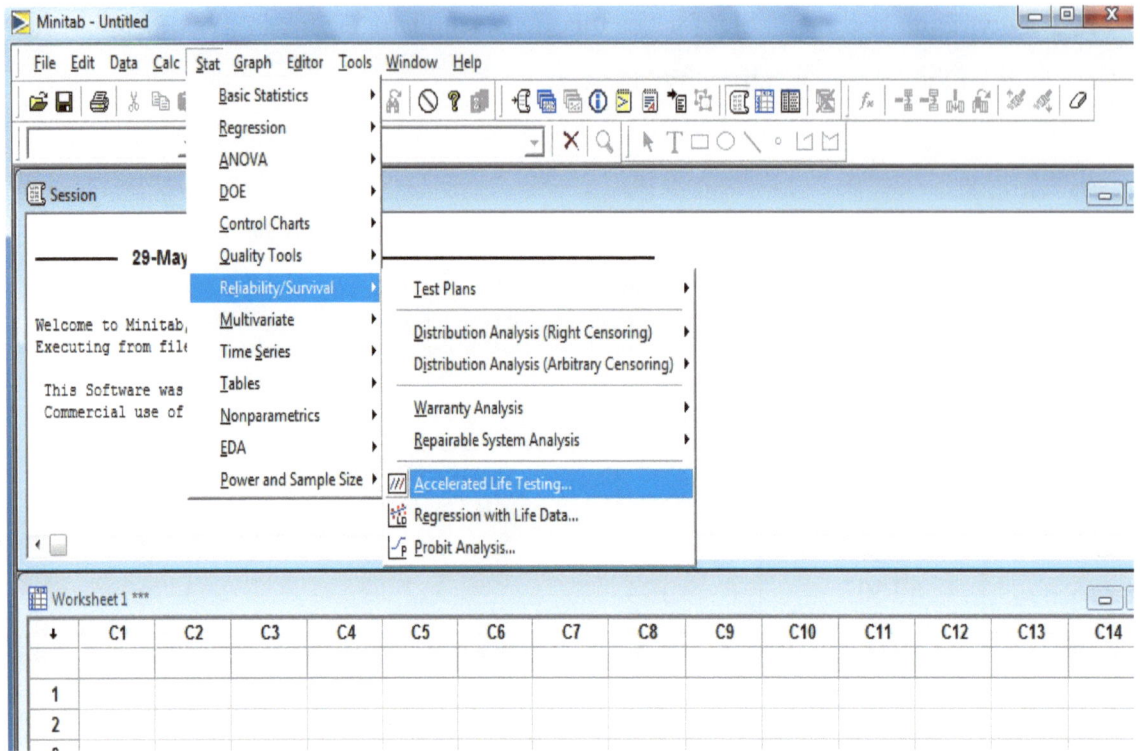

Fig. 6.1 Minitab Accelerated Life Testing window.

Step: 2 Accelerated Life Test level was selected in this step.

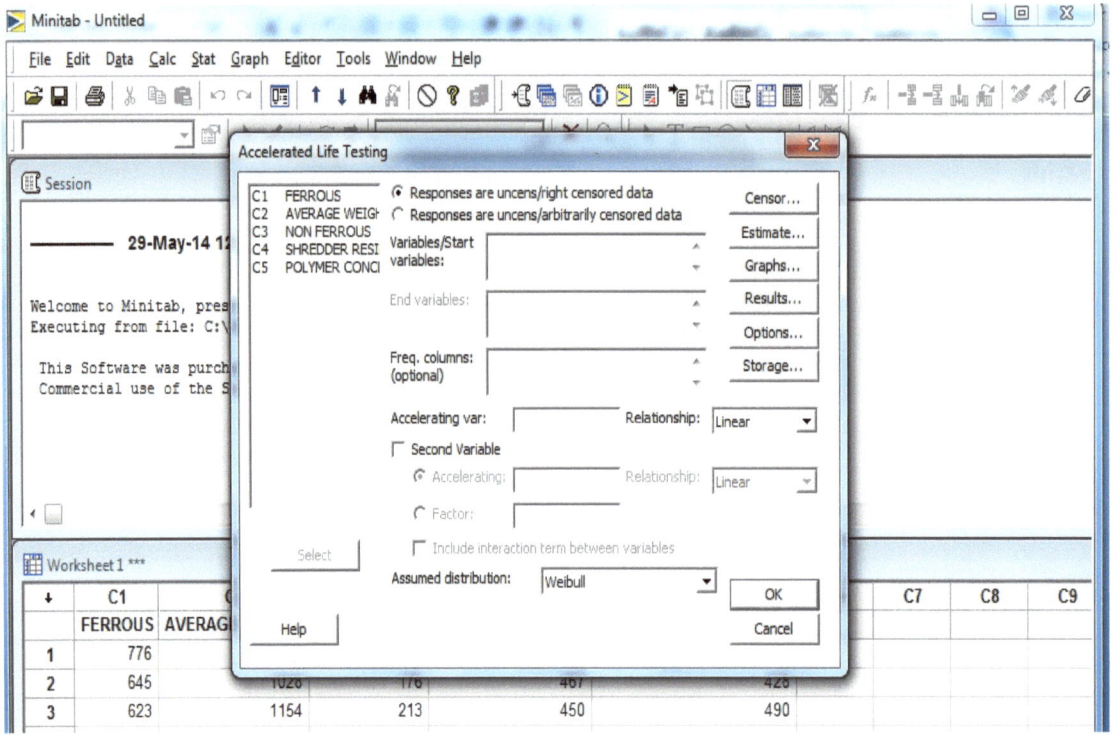

Fig. 6.2 Minitab Accelerated Life Testing level window.

Step: 3 Accelerated Life Testing Responses are uncens/right censored data was selected in this step.

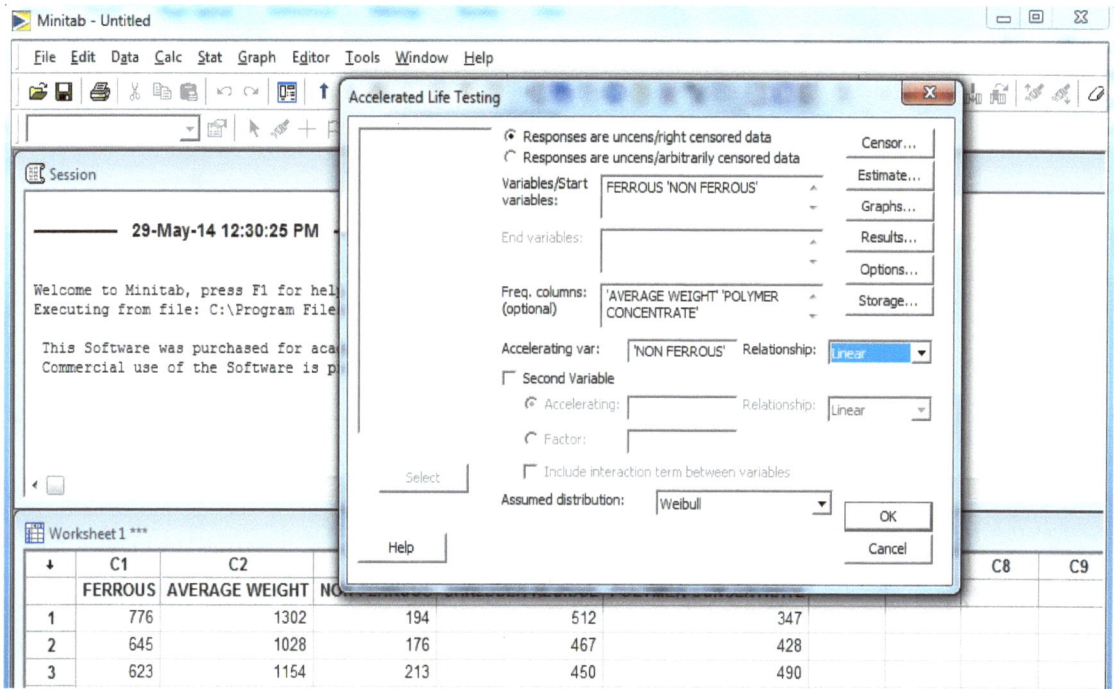

Fig. 6.3 Minitab Accelerated Life Testing Responses are uncens/right censored data window.

Step: 4 finally the Accelerated Life Test was Analyzed by using of this software.

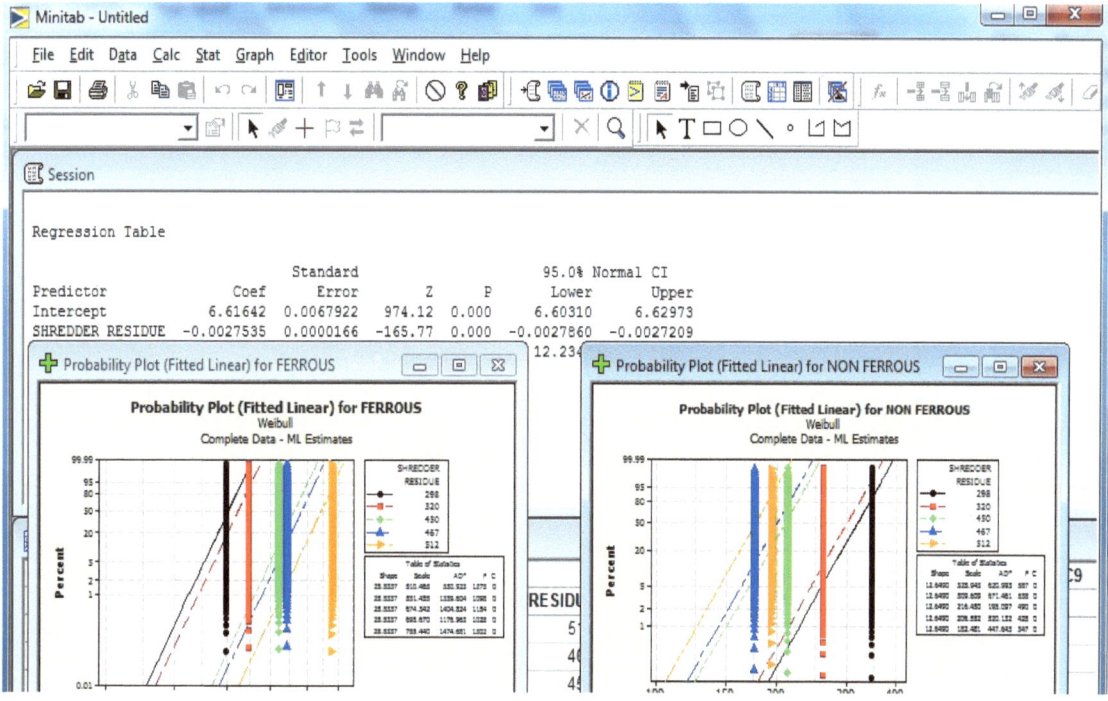

Fig. 6.4 Minitab Accelerated Life Testing Result window.

CHAPTER – 7

RESULTS AND DISCUSSION

7.1 Accelerated Life Testing

The experimental result for Accelerated Life Testing was given in the following graph.

Accelerated Life Testing: FERROUS versus SHREDDER RESIDUE

```
Response Variable: FERROUS
Frequency: AVERAGE WEIGHT

Censoring Information  Count
Uncensored value        5857

Estimation Method: Maximum Likelihood

Distribution:   Weibull

Relationship with accelerating variable(s):    Linear


Regression Table
```

		Standard			95.0% Normal CI	
Predictor	Coef	Error	Z	P	Lower	Upper
Intercept	5.68948	0.0022272	2554.49	0.000	5.68512	5.69385
SHREDDER RESIDUE	0.0018317	0.0000053	344.98	0.000	0.0018213	0.0018421
Shape	25.5337	0.273630			25.0030	26.0756

```
Log-Likelihood = -27883.024
```

Probability Plot (Fitted Linear) for FERROUS

```
Anderson-Darling (adjusted) Goodness-of-Fit
At each accelerating level

Level  Fitted Model
298        530.923
320       1339.604
450       1404.824
467       1176.963
512       1474.681
```

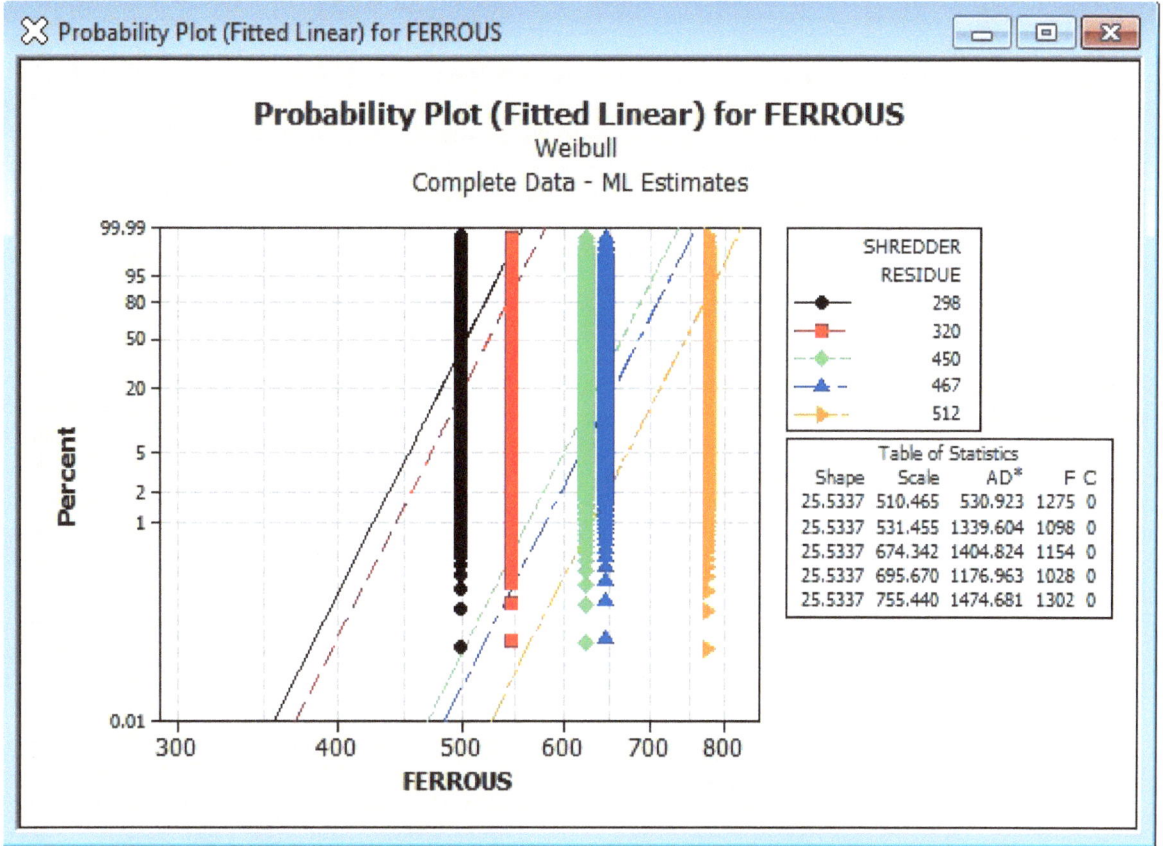

Probability Plot (Fitted Linear) for FERROUS
Weibull
Complete Data - ML Estimates

Legend — SHREDDER RESIDUE:
- 298
- 320
- 450
- 467
- 512

Table of Statistics

Shape	Scale	AD*	F	C
25.5337	510.465	530.923	1275	0
25.5337	531.455	1339.604	1098	0
25.5337	674.342	1404.824	1154	0
25.5337	695.670	1176.963	1028	0
25.5337	755.440	1474.681	1302	0

Accelerated Life Testing: NON FERROUS versus SHREDDER RESIDUE

```
Response Variable: NON FERROUS
Frequency: POLYMER CONCENTRATE

Censoring Information   Count
Uncensored value         2390

Estimation Method: Maximum Likelihood

Distribution:   Weibull

Relationship with accelerating variable(s):    Linear


Regression Table
```

		Standard			95.0% Normal CI	
Predictor	Coef	Error	Z	P	Lower	Upper
Intercept	6.61642	0.0067922	974.12	0.000	6.60310	6.62973
SHREDDER RESIDUE	-0.0027535	0.0000166	-165.77	0.000	-0.0027860	-0.0027209
Shape	12.6490	0.215316			12.2340	13.0782

```
Log-Likelihood = -10855.214
```

Probability Plot (Fitted Linear) for NON FERROUS

```
Anderson-Darling (adjusted) Goodness-of-Fit
At each accelerating level

         Fitted
Level    Model
298      620.993
```

```
320     671.461
450     195.097
467     520.132
512     447.643
```

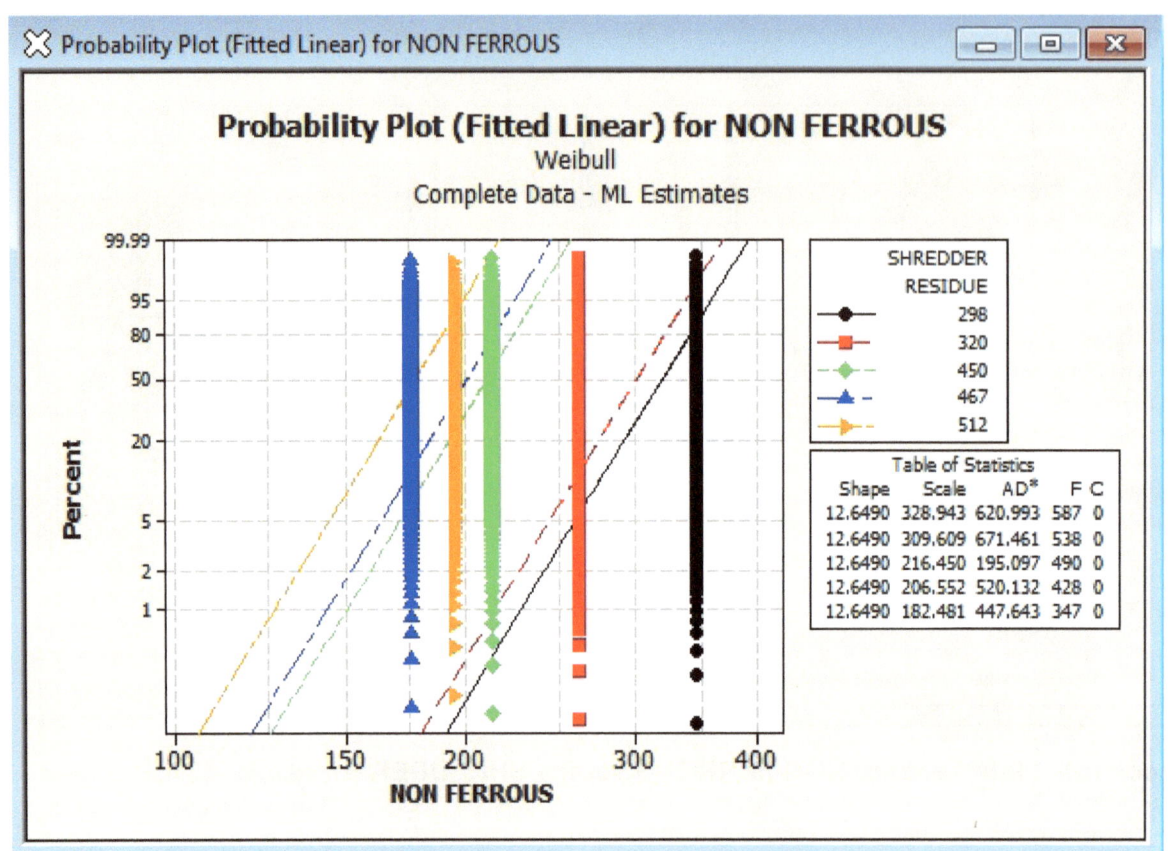

Relation Plot (Fitted Linear) for FERROUS, NON FERROUS

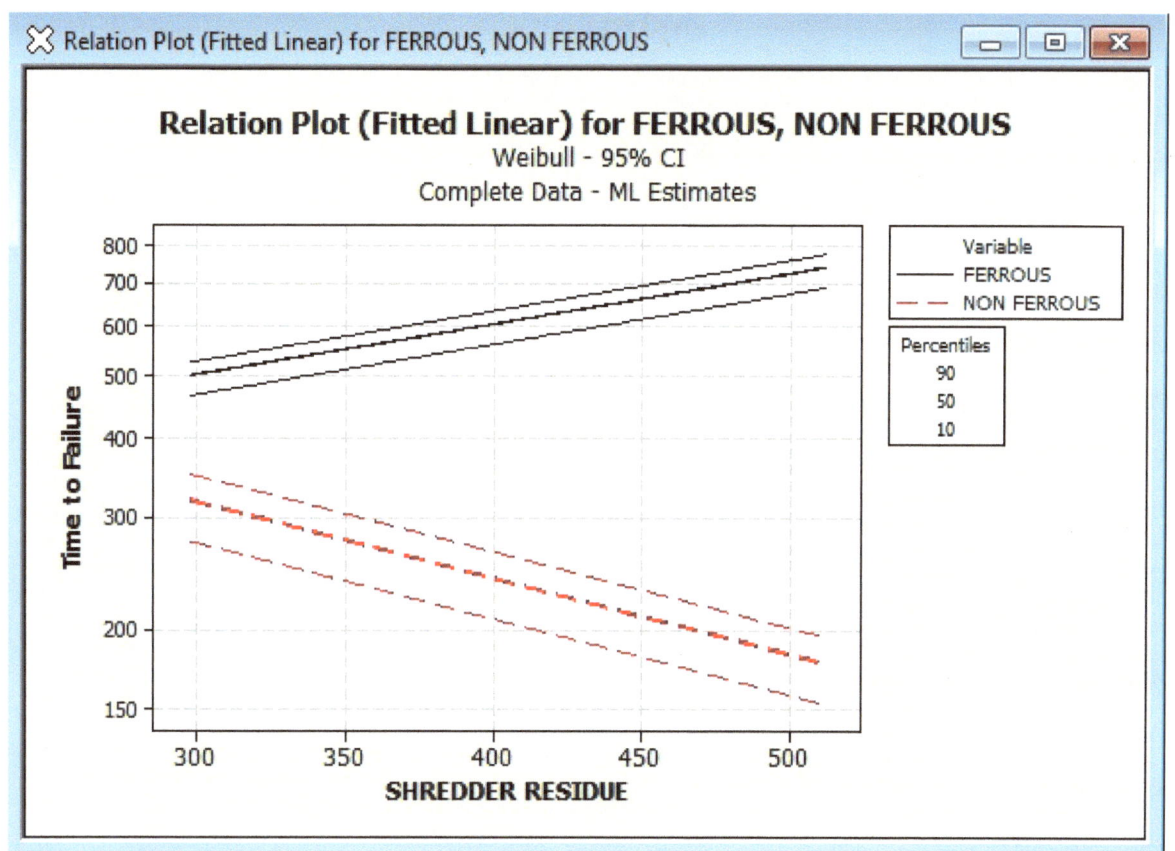

CHAPTER - 8

8.1 CONCLUSION

Conclusion of this project is an effective approach to improve recycling methods and creating new model for recycling. The modeling of the optimal ELVs treatment chain, in terms of best practices and best available techniques for both the improvement of the ELVs recycling rate and the production of cleaner streams of secondary raw materials. This study incorporates also the ASR energy recovery assessment along with material recycling; waste to energy is one of the industrial symbiosis opportunities to reduce the use of landfill disposal in ASR management. Finally, the logical sequence of treatments which are useful to completely remove metal from ASR, and to design and develop innovative treatment lines with an almost *"zero landfill"* target. The result shows the potential of the proposed methodology for improving customer service response in a dynamic world.

CHAPTER - 9

9.1 CREDITS

The paper work of this project is selected and presented in three International conferences.

- Dr.G.D.Sivakumar, S.Godwin barnabas, S.Anatharam entitled "Review on Automobile Material Recycling Management" paper presented in **Second International Conference on Advances in Industrial Engineering Applications** on January 6-8, 2014 at ANNA UNIVERSITY, CHENNAI.

- Dr.G.D.Sivakumar, S.Godwin barnabas, S.Anatharam entitled "Automobile Material Recycling Management" paper presented in **IEEE International Conference on Innovations in Engineering and Technology** on 21&22 March 2014 at KLN College of Engineering, Madurai.

- Dr.G.D.Sivakumar, S.Godwin barnabas, S.Anatharam entitled "Current Situation of End of Life Vehicle Treatment System In India" paper presented in **International Conference on Recent Trends in Engineering and Management** on 11&12 April 2014 at Indra Ganeshan College of Engineering, Trichy.

SECOND INTERNATIONAL CONFERENCE ON
ADVANCES IN INDUSTRIAL ENGINEERING APPLICATIONS DoIE

ICAIEA 2014

Certificate

This is to certify that

Anantharam. S

has participated and presented a paper entitled

" Literature Review on Automobile Material Recycling Management "

in the

Second International Conference on Advances in Industrial Engineering Applications held during January 6 - 8, 2014 at Anna University, Chennai, India.

Dr.P.Malliga
Convener

Dr.T.Ramesh Babu
Organising Secretary

Dr.T.Paul Robert
Chair

DEPARTMENT OF INDUSTRIAL ENGINEERING, ANNA UNIVERSITY, CHENNAI, INDIA

K. L. N. COLLEGE OF ENGINEERING

Madurai, Tamilnadu, India

2014 IEEE INTERNATIONAL CONFERENCE ON INNOVATIONS IN ENGINEERING AND TECHNOLOGY

ICIET'14

CERTIFICATE

This is to certify that Dr. / Prof. / Mr. / Ms. ......S...ANANTHA.RAM................................... has presented a paper titled

of...VELAMMAL...COLLEGE...OF...ENGG....&...TECH..........

.....AUTOMOBILE......MATERIAL......RECYCLING...MANAGEMENT...........................

....in "2014

IEEE International Conference on Innovations in Engineering and Technology (ICIET'14)"

organized by K. L. N. College of Engineering on 21st & 22nd, March 2014.

Dr. M. R. Thansekhar
Convener

Dr. N. Balaji
Convener

Dr. A.V. Ram Prasad
Principal

41

Indra Ganesan
COLLEGE OF ENGINEERING
Madurai Main Road (NH 45B), Manikandam, Tiruchirappalli – 620 012.

International Conference on
Recent Trends in Engineering and Management

ICRTEM 2014

Certificate

This is to certify that Mr/ Mrs/ Dr/ Miss ......ANANTHARAM.S....... has participated and presented a paper entitled ......CURRENT SITUATION OF END OF LIFE VEHICLE TREATMENT SYSTEM IN INDIA...... in the "International Conference on Recent Trends in Engineering and Management" held during april 11-12, 2014 at Indra Ganesan College Of Engineering, Trichy, Tamilnadu.

M. Mareeswaran
Convener

Dr. S. Bharathi Raja
Principal

Dr. G. Balakrishnan
Director

42

REFERENCES

[1] Meng, Song, "The Green Degree Assessment of the Automobile Reverse Logistics System" *IEEE International conference on Environmental Science and Information Application Technology, 2009, 112-115.*

[2] Hironobu, Susumu, "Inverse Factory for End-of-Life Cars-Complete Dismantling System", *IEEE International Journal of Advanced Engineering Sciences and Technologies, 2009.*

[3] Zhang, Wang and Chu, "Automotive Green Supply Chain Management Based on the RFID Technology" *IEEE 2010. 978-1-4244-6932-1/10.*

[4] Satoshi and Takeshi, "Graph-Based Information Modeling of Product-Process Interactions for Disassembly and Recycle Planning" *IEEE 2010. 978-1-4244-6931.*

[5] Sourkounis, "Dynamic Speed Flexible Drive System for Shredder-Plants with Highly Restricted Control Range" *IEEE 2008. 978-1-4244-1766-7.*

[6] Meng and Dong, "An Information System Management of Assessment of Disassembly and Recycle" *IEEE 2003. Q-7803-7743-5.*

[7] Zhang, Wang and Chu, "Automotive Recycling Information Management Based on the Internet of Things and RFID Technology" *IEEE 2010. 978-1-4244-6932.*

[8] Li-Shih and Shun-Lee, "Optimizing Disassembly and Recycling Process for EOL LCD-Type Products: A Heuristic Method" *IEEE 2007. 1521-334 X.*

[9] Rifer and Salazar, "Conceptualizing an Optimal Electronic Product Design and End-of-Life Management System" *IEEE 2007.1-4244=0861-X.*

[10] Rifer and Salazar, "Study on the Informationization Management for Auto-Remanufacturing Recovery System". *IEEE 2010. 978-1-4244-6932-1.*

[11] Shugang Zhu, "Recovering Copper from Spent Lithium ion Battery by a Mechanical Separation Process". *IEEE 978-1-4244 2011.*

[12] George, "Evaluation of Value Changes Between Different Phases of the Product Life-Cycle" *IEEE 9th International Conference on Computational Cybernetics July 2013.*

[13] Amy, Charles, "Using Fuzzy Cognitive Map for Evaluation of RFID-based Reverse Logistics Services" *IEEE International Conference on Systems, October 2009.*

[14] Gerrand, Kandlikar, "Is European end-of-life vehicle legislation living up to expectations? Assessing the impact of the ELV Directive on 'green' innovation and vehicle recovery" *IEEE 1-4244 (2009).*

[15] Muhammad Zameri, "End of life vehicles recovery: process description its impact and direction of research" *Journal of Mechanical June 2006.*

[16] Kee Mo Jeong, Lee, "Life Cycle Assessment on End of Life Vehicle Treatment System in Korea" *Journal of Industrial Engineering and Chemistry (2007).*

[17] Edward, Tracy, "A Design Framework for End of Life Vehicle Recovery" *CIRP International Conference on Life Cycle Engineering. (2006).*

[18] Bonollo, Moline, "Life Cycle Assessment in the Automotive Industry: Comparison between Aluminum and Cast Iron Cylinder Blocks" *Journal of metallurgical science and technology.*

[19] Jean Jacques, "Life Cycle Assessment Practices: Bench marking Selected European Automobile manufacturers" *International journal of product Life Cycle Management (2007).*

[20] Vishesh Kumar, John, "Sustainability of the automotive recycling infrastructure: review of current research and identification of future challenges" *International journal of sustainable manufacturing (2008).*

[21] Denis, Iorga, "New methods for recycling plastic materials from end of life vehicles" *ISSN (2011).*

[22] Muhamad Zameri, "Strategic Guidance Model for Product Development in Relation with Recycling Aspects for Automotive Products" *IEEE (2010).*

[23] Hendrix, Kevin, "Technologies for the Identification, Separation and Recycling of Automotive Plastics" *International Journal of Environmentally Conscious Design and Manufacturing (1996).*

[24] Pavel, "The Automobile Recycling Industry in North America" *IEEE 978-1-4244 2011.*

[25] Paulina, Kawa, "Remanufacturing in Automotive Industry: Challenges and Limitations" *Journal of Industrial Engineering and Management pp. 453-466 (2011)*

[26] Tawabini Pambid, "Trace Elements in the Leach ate of Automobile Scarp Shredder waste", *International Journal of Waste Management Vol. 11 p. 283-286 (1991)*

[27] Muhamad Zaneri Mat Samanm, "Design for End of Life Value Framework for Vehicles Design and Development Process" *Journal of Sustainable Development Vol. 5, March (2012).*

[28] Wilhelm, "Materials used in automobile manufacture - current state and perspectives" *Journal De Physique IV Vol.3, (1993).*

[29] Zsofia, "Best Available Technologies in End-of-Life Vehicles Recycling" *F2008-SC-011.*

[30] Miller, Wittebrood, "Recent development in aluminum alloys for the automotive industry" *Materials Science and Engineering A280 pp. 37-49 (2000).*

[31] Kun Yue, "Comparative analysis of scrap car recycling management policies", *Procedia Environmental Sciences, 16 (2012) pp 44 – 50.*

[32] Sameer Kumar, "System dynamics study of the Japanese automotive industry closed loop supply chain" *Journal of Manufacturing Technology Management Vol. 18 No. 2, 2007 pp. 115-138.*

[33] Pretz T., Mutz, "Metal Recovery in Automobile Recycling" *Acta Metallurgical Slovaca, 12, 2006, (456 - 462).*

[34] Nunes and David Bennett, "Green operations initiatives in the automotive industry an environmental reports analysis and benchmarking study" *An International Journal Vol.17 No. 3, 2010 pp. 396-420.*

[35] Apratul Chandra Shukla, S.G. Deshmukh, Arun Kanda "Environmentally responsive supply chains Learning's from the Indian auto sector" *Journal of Advances in Management Research Vol. 6 No 2, 2009 pp. 154-171.*

www.ingramcontent.com/pod-product-compliance
Lightning Source LLC
Chambersburg PA
CBHW050835180526
45159CB00004B/1911